Introduction

Photovoltaic Power System

Assessment of the Site

Orientation of the Solar Arrays and Shadow

Space

Tilt Angle

Conditions of the mount surface

Calculating the Size of the PV System

Choice of Equipment

Inverters

Racking Systems

Energy storage

Surge Protection

Interviewing a Contractor

General Tips on PV System Installation and Operation

Introduction

For society to fully be able to satisfy its energy d1emands, three major factors must be considered; security of energy supply, impact of the energy supply on the climate and environment and economic cost of the energy. These three factors raise important questions:

- Can we steadily generate and supply the energy we need? Shortage of energy is undesirable. We are dependent on the availability of energy to carry out our day to day activities (heating, transportation, communication, lighting, entertainment, cooking, and power equipment) and to maintain the standard of living in the society. A reliable, steady supply of energy is essential and must be guaranteed to reduce dependence on energy importation.
- Can we generate energy is a manner that has a less detrimental impact on the environment? Fossil fuels serve as the primary source of energy globally but has major negative environmental impact like CO_2 emissions which is associated with the increased greenhouse gas (GHG) levels in the atmosphere and contributes to global warming. There is a drive to reduce the level of CO_2 emissions by utilizing a mix of energy sources that do not involve combustion of fossil fuels.

- Can we achieve cheaper energy costs? The various sources of energy has a corresponding cost due to the difference in method and technology. The cost of energy production determines the final cost society will pay for energy devoid of subsidies. Keeping energy costs low will keep the energy mix competitive

No singular source of energy can serve as the answer to these questions. Fossil based energy production is relatively cheaper but detrimental to the environment and may involve a country relying on importation of the fuel required just as the rate of depletion of the existing reserves is fast paced. Renewable sources are sources of energy that are continuously available in the environment like wind, solar and hydro are cheap sources of energy, supply is intermittent but has minimal impact on the climate and environment. Renewable energy sources are infinite and served mankind for years. If properly harnessed, renewables have the potential to replace or reduce dependence on fossil fuels as the primary source of global energy supply.

Therefore, the three constraints must be optimized for us to effectively balance the three factors under consideration.

The sun is a clean, environmentally friendly, abundant and free renewable resource that can be harnessed to produce energy. The total sunlight falling on many countries of the world is enough to satisfy her energy demands several times over.

Solar PV systems are modular and can be utilized for both off-grid and grid connected installations. Photovoltaics are adaptable for use in small-scale generation for use on-site in homes, offices or industries. Other advantages of PV systems include the low operations, maintenance costs and complexity.

With the level of investments into research, PV is expected to become more economically viable for a wider range of applications in the coming years.

Photovoltaic Power System

Photovoltaic is primarily a combination of two words; photo and voltaic meaning light and electricity respectively. A PV system converts sunlight received from the sun into electric current. Once the sunlight strikes the solar cell, electrons are released from its atoms and electrons flow through the semiconducting material to product electricity. A number of solar cells connected together form a Solar panel which are then connected together to form an Array (Solar Array). Majority of the solar cells in the market today are made from Silicon (Si). The semiconducting property of the Silicon makes it especially useful for this purpose as its electrical conductivity lies between that of insulators and metals. Each Silicon atom has four valence electrons that have a strong bond with their atoms. However, when photons of light strike Silicon, the photon's energy is absorbed by the valence electrons which weaken the strong bond and releases the valence electrons from their bonds and transform into mobile charges leaving behind a hole where an electron could reside but is currently empty. For the negatively charged electron, the hole is seen as a positive charge carrier and contributes to the current. The Photovoltaic cell has built-in electric field that provides the required voltage to push the current through an electrical equipment.

Silicon cells have different types of atomic structures; mono-crystalline, multi-crystalline and amorphous.

Other materials that may be used as Semiconductors include Cadmium Telluride, Copper Indium Diselenide.

The energy output of the sun daily is estimated to be about 8.3×10^{25} kWh while the earth receives about 4.2×10^{15} kWh. The radiation can be converted into electricity by making use of PV devices. The amount of solar energy available at a particular location can be estimated using the level of irradiation (also referred to as insolation) while the amount of solar power can be estimated using the irradiance of that location. These two values are required for the design of a PV system and can be gotten from a repository of meteorological data taken around that location.

A PV array is made up of solar modules which contain solar cells. Solar modules consist of several solar PV cells built and sealed together and usually between 10 and 320 watts.

Solar PV systems may either be grid connected (connected to the electric utility grid) or off grid (not connected to the grid but to a stand-alone / battery system that stores the energy for use).

Assessment of the Site

Sunlight is the equivalent of fuel in PV system as natural gas would be to a gas turbine. Therefore, it is essential to have an estimate of the solar resources available and the day / seasonal variations at a particular location to effectively approach the system design.

One of the first things to do when planning a Solar PV installation is to do a thorough assessment of the location with the objective of determining the best location to install the solar arrays. There are two basic choices; a roof-top mount or a ground mount.

Orientation of the Solar Arrays and Shadow

To derive maximum energy output, the solar arrays should be tilted facing the direction where the sun is most visible, perpendicular to the sun rays. One of the biggest causes of low performance in Solar PV installations is the poor orientation of the solar arrays. Therefore, the site assessment and orientation of the sun has to be done carefully.

Also, care should be taken so that the quality of irradiation incident on the solar arrays is not compromised by any obstruction from other buildings, smoke, trees or any objects that can serve as a source of obstruction that will likely result in a shadow on the arrays. A particular area can be unshaded at some points during the day and be shaded at other points of the day.

There are solar shading instruments such as Solar Pathfinder and Solmetric Suneye that can assist you to identify potential sources of shadows and an estimate of time the shadow will affect the solar arrays, resultantly helping to determine the optimal site to install the arrays.

Space

The amount of energy captured is dependent on the number of solar arrays that are exposed to sunlight. All things being equal, an increase in the area of solar cells being used will result in an increase in the energy output. The space available for the installation will determine the quantity of solar modules to be used and the amount of space that would be utilised. You have to determine whether the property has enough free space to house the solar arrays. Usually, solar arrays are mounted on roof top. Does the roof top have enough available space to accommodate the solar arrays? Other places that could be considered are pole mounts, wall mounts, ground mounts or as part of a shade.

Focusing or concentrating the sunlight with a mirror and lenses would reduce the number and invariably cost of PVs required. It has also been proven over time that the efficiency of PV cells improve with concentrated sunlight.

Ultimately, if there exists a space constraint you may consider installing PV modules with higher efficiency.

Tilt Angle

The tilt angle is the angle between the solar panel and the horizontal plane. Solar arrays are most efficient when they are installed perpendicular to sun rays. The amount of sunlight falling on a solar array is directly proportional to the amount of electricity that will be produced. Usually, the tilt angle for optimal annual energy output is the location's latitude plus 15 degrees in winter or minus 15 degrees in summer.

The rule of thumb states that for locations in the Northern Hemisphere, solar arrays should face true south while for locations in the Southern Hemisphere, the solar arrays should face true north. However, it is impossible to have a tilt angle that will be optimal at all times since the sun's angle changes daily from east to west. A si1ngle or dual axis tracking device may be installed to improve the performance of the solar arrays but a computation of the estimated increase in output over time and the cost of the tracking device should be done to weigh the payback period and determine whether investing in the device would be viable. There exists manually operated tracking designs that allows the tile angle to be adjusted manually.

Conditions of the mount surface

The mount surface should be thoroughly examined to identify any underlying issues such as leaks or cracks so that repairs or replacements can be effected before the arrays are installed. Replacing a roof not long after a PV system is installed would not be a cost-effective decision. The solar arrays will add some weight to the surface, therefore the ability of the surface to carry such additional weight should be verified.

Solar arrays are negatively impacted by excessive heat which may result from heat transferred from the roof. Reverse bias (a cell drawing current from other cells) resulting from partial shading of the solar module. Partial shading may be caused by protrusions on the roof, trees, dead animals and waste or electric poles imposing a shade on the solar arrays. To prevent reverse bias and partial shading, suitable spacing, ventilation and avoidance of shaded modules should be factored in while installing the solar arrays.

Calculating the Size of the PV System

Upon completing the evaluation of the proposed location, the next thing to do is to design a solar array that would satisfy the objective(s) of the project with suitable thoughts and consideration for any constraints that has been uncovered.

Prior to determining the size of the PV system, it is important to define the underlying goals for embarking on the project. This will go a long way to guide the decision-making process and the choices that would be made along the way. Why are you doing this

- To make money from supplying of electricity to the grid
- To be on the path to or become self-sufficient and halt using the power from the grid resultantly minimizing energy costs
- To add your quota to the quest to cut carbon emissions and make the environment cleaner
- To create a backup power source in the event of power outages

To adequately size the system, estimate the energy needs of the location in kWh. To do this, you can refer to the previous electric bills and calculating an average of the last three months. Alternatively, you can make a list of all appliances including the power consumptions of each of them in Watts and how many hours they are in use weekly. You can then use this to calculate the annual energy requirements. Bear in mind that the PV system must produce enough energy to cover the load requirements. The energy supply from a PV system is dependent on the choice of PV modules, inverter, the arrangement and placement of the modules and the weather conditions of the location.

Next, the solar insolation of the location (watts per square meter)

Adjustments of 15-20% should be made for unavoidable system losses that may result from shading, inverter efficiency, wet PV cells, snow, wiring and other sources.

For instance, to meet energy requirements of 10,000 kWh per year in Las Vegas, Nevada, USA. Assuming system losses of 20% and solar array efficiency of 15%.

- The solar insolation for the Las Vegas is an average of 6.41 kWh/m²/day at a tilt angle of 21° (latitude of 36° less 15)

- Rating of solar arrays required is 10,000kWh / 6.41 kWh/m² /day x 365 days x 0.80 = 5.3 Kw

- Quantity of solar modules using a 300 Watt (0.3kW) solar module with 72 cells per module (cell layout of 12 x 6) and dimension of 1.968 m x 0.987 m. The area of one 300 watt solar panel is 1.94 m² = 5.3kW/0.3kW = 17.67 which would be rounded up to 18 solar modules

- The solar arrays can be arranged in a 2 x 9 or a 3 x 6 formation depending on the available space for the installation.

- The area of one solar module is 1.94 m². Therefore, the available space for the installation should be 1.94 m² x 18 = 34.92 m²

This calculation can also be done using software or spreadsheets (some freely available online) but the results are approximately the same depending on the assumptions made.

Choice of Equipment

There exists a wide variety of commercially available components of a solar array. Generally, the choice of equipment should be based on the following

- Track record of the manufacturer
- Quality of the product: the differentiating factor between two components of a PV system can be say efficiency of the Solar Modules. Efficiency goes a long way in the maximum energy derived from Solar arrays. it is best to conduct a market survey and make comparisons of different products based on information provided by manufacturers.
- Price
- Adherence to International Standards
- Warranty
- Rated Power: Rated Power is the maximum power a panel can produce with an input of 1000 Watts/m^2 with the modules operating at ambient temperature (25°c). of sunlight Usually, residential PV systems make use of solar modules with power outputs of 100-300 Watts

Some of the equipment required in a PV system include

- Solar PV modules
- Racking System

- Inverter
- Surge Protection
- Meters
- Battery

Inverters is one of the major components of a PV system. PV arrays produce Direct current electricity while it is necessary to use Alternating Current to work with certain electric loads. While Direct current is a useful form of electricity, it is not suitable for every application.

Inverters transform the direct current from the PV modules or battery to alternating current while ensuring that the frequency of the AC cycles is 60 cycles per second. Inverters regulate the output voltage by reducing fluctuations.

The inverters Direct Current Voltage should match the nominal voltage of the solar array dependent on whether the configuration includes battery or not. Inverters usually work at partial loads for most of their useful life implying that the efficiency of an inverter at partial load should be high. The load to be served will determine the sort of inverter to use.

- Square wave inverters for resistive loads

- Modified sine wave inverters
- Sine wave inverters

An inverter with high reliability and efficiency would be required in to ensure that the PV system operates at the maximum power possible. It is best to locate the inverter in a cool area

Racking Systems are pre-engineered systems made up of aluminium or steel that provide support to hold the Solar arrays together in place securely enough to resist winds, rainfall, theft and other forces that may cause it to trip over. Good choice of racking system also makes the installation beautiful to the eye and ensure proper airflow around the PV system.

- Tracking the sunlight: The racking system could feature a tracking system that would allow the solar arrays follow the motion of the sun and derive maximum input. Generally, tracking would increase the amount of energy input but the cost-effectiveness has to be determined before investing in the device.
- Adjust the PV system: The racking system allows you to adjust the angle of the PV modules depending on the season or time of day.

The racking system should be such that it is still easily accessible during periods of repairs and maintenance.

Energy storage

Without a form of storage, a PV system will only generate current when there is light to kick-start reactions and as such are unproductive most of the time. During the night or cloudy days, the solar arrays may not work at all or produce as much energy as it does during the periods where there is more light (5-25% of the maximum output). Therefore, it is not uncommon to have PV systems where storage systems (usually rechargeable batteries) are incorporated though the configuration is relatively more expensive when compared to a configuration without energy storage. Battery will maintain power to electric equipment when there is no supply from the PV system. This extends the availability of electricity to periods when there is no light to power the PV system.

Also, batteries affect the complexity and cost of the installation and require regular maintenance checks.

The choice of battery will be determined by the location, ambient temperature range and intended use of the installation though the most commonly used is the Lead Acid Battery. Other battery options include lithium-ion (Li-ion) and sodium-sulphur batteries (NaS), Pocket-Plate Nickel Cadmium, Nickel metal hydride.

The reserve capacity of the battery should be substantial for the PV system to supply energy during periods of low or no sunlight (cloudy weather). Conventionally, for residential applications, reserve capacity should be about 5 days.

The capacity of a battery can be computed by multiplying the daily direct current energy requirement of the PV system by the number reserve capacity days with adjustments made for system losses. For long lasting operational life, batteries should only work up to a maximum of 80% of its capacity. This implies that the capacity of a battery is determined by multiplying the required capacity by a factor of 1.25.

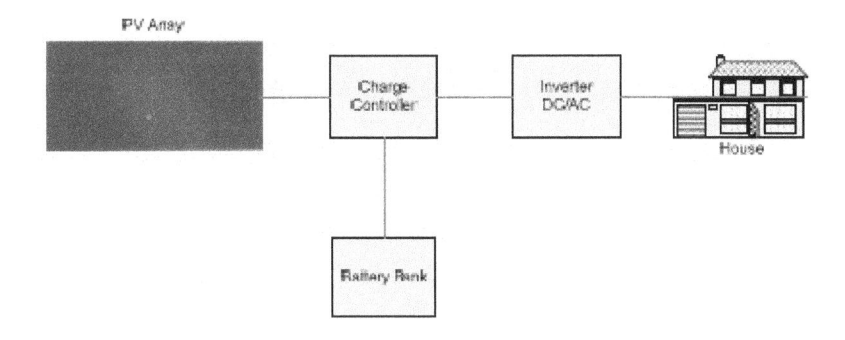

Surge Protection

An increase in voltage beyond the design limits (surge) can damage the PV system and connected equipment. A surge protector provide protection to the system from power surges that could possibly occur during the operation and from lightning.

Interviewing a Contractor

Below is a collection of questions to ask a vendor you are considering contracting for a Solar Project.

- Does the contractor have the requisite credentials to prove that they can handle the project?
- Does the contractor own or have access to the required tools?
- Does the contractor have references to validate that previous project experience?
- What is the payment schedule (upfront payment or payment after milestone)?
- Is the PV system scalable? Can more solar arrays be added in the future to increase the energy output?
- What are the details of the warranty being offered?
- How long will the design and installation take?
- What are the possible impacts of the installation on the existing structure (will it cause roof leaks or constitute a heavy weight)
- What is the payback period of the investment?
  Payback period is the amount of time it will take for the net savings as a result of the project will outweigh the initial outlay on the project.

This will go a long way in determining whether the project makes financial sense.

General Tips on PV System Installation and Operation

- The PV modules are sometimes coated with glass to protect the solar cells from oxygen, humidity, dust and rain. Therefore, please handle with care to prevent breakage.
- The right polarity should be maintained at all times for all connections. Conventionally, use a red wire for all positive connections and a black wire for negative connections.
- Please earth the solar arrays by placing metal or earth rods. If adequately and properly earthed, lightning strikes would not affect the solar arrays.
- Clean the exposed portion of the modules regularly to wipe off dust
- Blocking diodes should be fixed in the solar arrays to ensure reverse current flows into the modules are prevented. Reverse current flows can cause energy losses and damage the solar modules.
- Battery should undergo regular checks and electrolyte should be refilled.
- The wire sizes required for DC power are larger than for AC power at same voltage.

- The array wiring must be done in such a way that it can withstand elevated temperature.
- Every component must be carefully selected to ensure that the overall system is able to operate optimally.

CPSIA information can be obtained
at www.ICGtesting.com
Printed in the USA
BVHW020229091222
653821BV00009B/33